BALANCED BREAKTHROUGH RESPONSIBLE TECHNOLOGY

A Net-Zero Guide for Promoting Sustainability
and Human Well-Being for All

BALAKARTHIK BASKARAN

authorHOUSE

AuthorHouse™ UK
1663 Liberty Drive
Bloomington, IN 47403 USA
www.authorhouse.co.uk
Phone: UK TFN: 0800 0148641 (Toll Free inside the UK)
  UK Local: (02) 0369 56322 (+44 20 3695 6322 from outside the UK)

© 2023 Balakarthik Baskaran. All rights reserved.

No part of this book may be reproduced, stored in a retrieval system, or transmitted by any means without the written permission of the author.

Published by AuthorHouse 03/22/2023

ISBN: 979-8-8230-8143-6 (sc)
ISBN: 979-8-8230-8142-9 (e)

Print information available on the last page.

Any people depicted in stock imagery provided by Getty Images are models, and such images are being used for illustrative purposes only.
Certain stock imagery © Getty Images.

This book is printed on acid-free paper.

Because of the dynamic nature of the Internet, any web addresses or links contained in this book may have changed since publication and may no longer be valid. The views expressed in this work are solely those of the author and do not necessarily reflect the views of the publisher, and the publisher hereby disclaims any responsibility for them.

*For my parents Baskaran and Thenmozhi,
and my in-laws Jegaperumal & Ananthi:
thank you for your good principles,
guidance, love, and support*

*For my lovable wife Priya:
thank you for your love and support*

*For my dear prince Nikhil, little princess Tharika,
lovable sister Priyanka, brother Kumar,
brother-in-law Guhan, and other relatives:
thank you for your love and support,
and always try to use technology responsibly
and to leverage the magic of technology
for self, people's betterment,
and the environment.*

*For the scientists, innovators, technologists, activists,
and common humankind who are leading the way*

CONTENTS

Chapter 1	Breathing the Oxygen of Digital Technologies	1
Chapter 2	The Blessings and Detriments of Digital Technologies	9
Chapter 3	The Rise of Balanced-Breakthrough Responsible Digital Technology	16
Chapter 4	The 2BReDiT Framework	27
Chapter 5	Principles of Frugal Design	36
Chapter 6	Green Consumerism as a Business Model	46
Chapter 7	Digital Greening from the Laws of Nature	56
Chapter 8	The Missing Element	65
Chapter 9	Responsible Technology: Forerunners and Present Action	69

References 77
Acknowledgements 83
About the Author 85

Breathing the Oxygen of Digital Technologies

Digital technology combines the art of captivating human digitalism with a focus on aiding humans to meet an objective with ease and supplemental help. The recent improvements and openness in technology have paved the way for endless supplemental technologies and a few critical dimensions, along with problems that can impact the environment and human cognitive states and boundaries.

The current modern digital world has empowered humankind with multiple digital services, including millions of songs, videos, WhatsApp messages, calls, and so on. However, many of the people who have these digital services at their fingertips are not cognizant of the fact that they are leaving a significant carbon footprint. Can you believe that a tiny message you send is emitting greenhouse gases, which in turn are heating up the world and making the glaciers melt? Yes, it's a plain fact. A massive amount of people messaging is already doing that. Research suggests that "to avoid catastrophic consequences from climate change, all sectors of the global economy, including *Information Communication Technology (ICT)*, must keep their greenhouse gas (GHG) emissions in line with the Paris Agreement" (Freitag et al.

2022). The Paris Agreement, often referred to as the Paris Accords or the Paris Climate Accords, is an international treaty on climate change. Adopted in 2015, the agreement covers climate change mitigation, adaptation, and finance.

ICT's share of global GHG emissions is at 1.8%–2.8%. ICT's carbon footprint is the amount of carbon generated by the information and communication technology sector. Just as an example, it is estimated that a single email accounts for 4 grams of CO_2e issued. An email with an attachment account for 50 grams of CO_2e (source: ictfootprint.eu). Digital technologies have created a vast amount of unmanaged or unsupervised content and electronic waste, which has a direct impact on human well-being, climate, and the environment.

However, digital technology has been a saviour as well. In a recent post, the World Economic Forum announced that "if brought to scale, digital technologies could reduce emissions by 20% by 2050 in the three highest-emitting sectors: energy, materials, and mobility." The recent development in technologies like digital twins, ethical AI, and strategic intelligence, as well as the harnessing of digital technologies to manage distributed energy sources in real-time collectively and so on, makes digital a power of good.

Key Global Problems

Following are some of the recent problems with respect to greenhouse gases.

- "The current emissions from computing are about 2% of the world's total but are projected to rise steeply over the next two decades. By 2040, emissions from computing alone will be more than half the emissions level acceptable to keep global warming below 1.5° C. This growth in computing emissions is unsustainable; it would make it virtually impossible to meet the emissions warming limit".
- "The emissions from the production of computing devices far exceed the emissions from operating them, so even if devices are more energy efficient, producing more of them will worsen the emissions problem. Therefore, we must extend the useful life of our computing devices."
- "In yet another ominous climate change warning, atmospheric levels of the three main greenhouse gases—carbon dioxide, methane, and nitrous oxide—all reached new record highs in 2021", according to a 2022 report from the World Meteorological Organization (WMO) (source: https://public.wmo.int/).
- "WMO's Greenhouse Gas Bulletin reported the biggest year-on-year jump in methane concentrations in 2021 since systematic measurements began nearly forty years ago. The reason for this exceptional increase is not clear but seems to be a result of both biological and human-induced processes."

- The latest scientific report by the IPCC finds changes in the earth's climate in every region and across the whole climate system. Many changes are unprecedented in thousands if not hundreds of thousands of years. Some, such as continued sea-level rise, are irreversible over hundreds to thousands of years.
- Countries export much of the harm created by their greenhouse gas (GHG) emissions because the earth's atmosphere intermixes globally. Yet the extent to which this leads to inequity between GHG emitters and those impacted by the resulting climate change depends on the distribution of climate vulnerability.

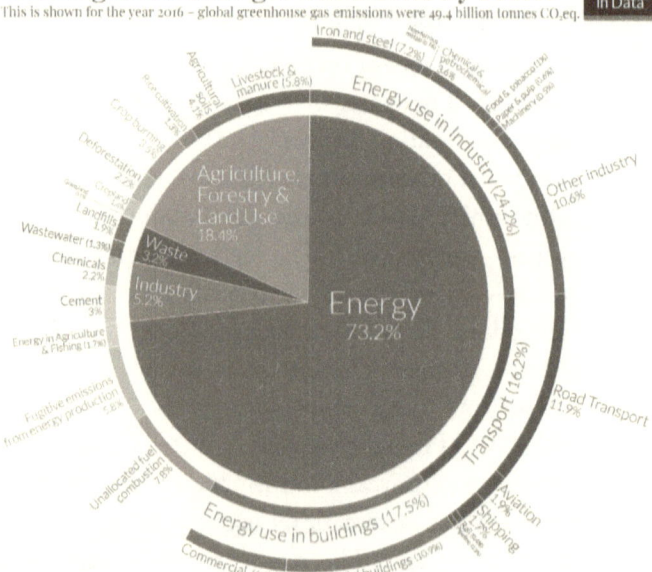

The Nexus of the Greenhouse Effect

The greenhouse gases emitted into the atmosphere stay there for a long time, resulting in increased carbon dioxide which will impact life on earth and trap more heat in the atmosphere. However, the goal is to achieve near-zero greenhouse gas (GCG) emissions and remove the gases that are already built up in the atmosphere—net negative emissions.

The combination of all the different greenhouse gases is known as "carbon dioxide equivalents (CO_2e)". The greenhouse gas emissions have also increased based on the burning of fossil fuels such as coal by humans.

However, the greenhouse effect is for good and for the benefit of the planet. How so? Sunlight, for example, passes through greenhouse gases without getting absorbed and warms up the planet. The planet would be far too cold without this effect. But the additional GHG emissions by human activity make it difficult for the planet to thrive and create biodiversity, well-being, and other environmental implications.

Controlling Climate Disasters with Clean Energy

According to *How to Avoid a Climate Disaster* by Bill Gates, 52 million tons of greenhouse gases are typically added to the atmosphere by the world every year, with an indirect impact on energy poverty. For example, about a billion people don't have reliable access to electricity. Studies suggest that a country's per capita income and the amount of electricity go together, along with increased investment in biotechnology and information technology.

Clean energy is one of the key solutions to overcoming many greenhouse problems. Clean energy, or renewable energy, is energy collected from renewable resources that are naturally replenished on a human timescale. It includes sources such as sunlight, wind, the movement of water, and geothermal heat. Although most renewable energy sources are sustainable, some are not. Clean energy production is a completely different ball game. Many governments came together in 2015 to create innovation missions for renewable or clean energy research. When the pandemic occurred in 2020, less greenhouse gases were emitted, but this decrease came with devastating costs and impacts on many people's lives and incomes. Post-pandemic, the resumption of a modern consumer lifestyle has brought an uptick in the emission of greenhouse gases, a trend that could increase. Clean energy innovations, research, and implementations, however, should continue to reduce the carbon footprint and other impacts on the environment.

The world is leading towards net zero, which means cutting greenhouse gas emissions to as close to zero as possible, with any remaining emissions re-absorbed from the atmosphere, for instance by oceans and forests.

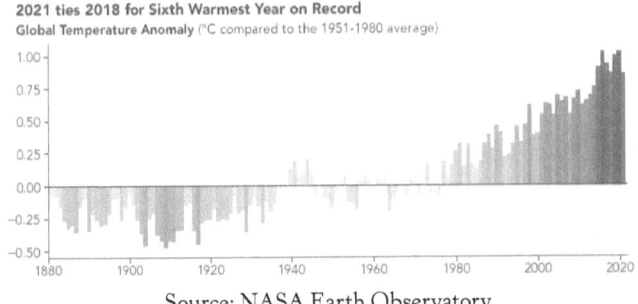

Source: NASA Earth Observatory

Science shows clearly that in order to avert the worst impacts of climate change and preserve a liveable planet, global temperature increase needs to be limited to 1.5 °C above pre-industrial levels. Currently, the earth is already about 1.1 °C warmer than it was in the late 1800s, and emissions continue to rise. To keep global warming to no more than 1.5 °C—as called for in the Paris Agreement—emissions need to be reduced by 45% by 2030 and reach net zero by 2050

The Goal of Creating a Sustainable Future

To summarize, technology has existed for multiple centuries and decades, helping humankind to be efficient in their lives and work. However, the myriad of technologies and accelerating adoption have left out some global environmental issues that humankind is facing as a challenge. Yes, you heard it right: technology is part of the problem, and surprisingly, technology is also the solution.

Responsible digital technologies can be a prime factor in achieving near-zero goals related to energy poverty, geopolitical tensions, future pandemic preparation and response, supply-chain disruptions, climate change, the rise of AI, and a growing risk of an inflation-driven recession that is leading to a period of extreme disruption.

Digital technologies have also been part of vast greenhouse gas emissions. I consider balance theory to be one of the core philosophies for creating responsible digital technology, which I will discuss through a few frameworks, principles, and considerations in the following chapters. Frugality can be woven into the software industry at all levels and related societies and works. Digital technology can be responsibly practiced, help us move towards near-zero goals, and be extended for multiple contexts while still maintaining physical and mental well-being, the environment, society, and governance (ESG).

The Blessings and Detriments of Digital Technologies

2

Information and communication technology (ICT) solutions have a calculated potential to reduce global greenhouse gas emissions by up to 15%, as per a report by Ericsson. Technologies in action today are improving lives, say a car taxi application or internet banking, and there are tens of billions of internet-connected devices around the world. While these technologies fulfill their purpose, they create problems for the end consumers and the planet. On the other hand, while machines are intended to be an aid to humans, technologies like artificial intelligence have the power of questioning humankind and disrupting the environment, climate, and human well-being. Digital technology is used in multiple areas of life such as business, education, banking, media, healthcare, transportation, communication, and software.

It is vital to evaluate the ebbs and flows of digital technology and apply the right technology for the right need. Some of the pros and cons of digital technologies are compiled here for a rational judgement of how digital technology can be leveraged.

Detriments Caused by Digital Technologies

Cyber attack. As an example, the wide-ranging use of smart technologies is raising global agricultural production, but international researchers at Flinders University have warned this digital-age phenomenon could reap a crop of another kind: cybersecurity attacks. Smart sensors and systems are used to monitor crops, plants, the environment, water, soil moisture, and diseases.

> "The transformation to digital agriculture would improve the quality and quantity of food for the ever-increasing human population, which is forecast to reach 10.9 billion by 2100,". This progress in production, genetic modification for drought-resistant crops, and other technologies is prone to cyber-attack—particularly if the ag-tech sector doesn't take adequate precautions like other corporate or defense sectors.

Data security. With the advent of new technologies and accelerated digital adoption, data security is still observed as a great concern. Cyber crimes are more prevalent with the breach of data and enterprises across the world are devising stronger mechanisms day by day to control data breaches.

Doomscrolling and social isolation. *Doomscrolling* or *doomsurfing* is the act of devoting an excessive amount of screen time to the absorption of negative news. Increased consumption of predominantly negative news may result in harmful psychophysiological responses in some cases. In the social world, face-to-face communication has become rare; communication is only by means of the internet.

Impact on jobs. Digital also has affected jobs in that people available through remote work may replace local jobs. Automation and artificial intelligence have had effects on jobs as some manual tasks are automated or replaced.

Digital technology addiction. Video games, social platforms, chats, singing apps, and other websites have proven to be addictive, and many people use them for hours and days.

Intellectual property violation. The digital world offers countless ways to manipulate digital media, morph photos and videos, violate copyrights, and practise plagiarism.

Virtual living. Real life and direct experiences are diminishing. Many people are attending video weddings, concerts, birthday parties, and other events, living through the prism of digital technology.

Physical health. Musculoskeletal issues, digital eye strain, disrupted sleep, physical inactivity, psychological issues, and hearing loss are some of the negative impacts of digital technology on health.

Cryptocurrency. The emissions from blockchain have generated a lot of attention in recent years because of the rise of cryptocurrencies such as Bitcoin. Cryptocurrency uses a blockchain to add transactions to a digital ledger. The "miners" solve complicated computer problems to confirm blocks of transactions and are rewarded with digital coins. However, the computing power needed to solve these problems is extremely energy-intensive.

As new digital technologies are created and leveraged, we will undoubtedly become aware of more negative impacts.

Blessings of Digital Technologies

Connectivity and Communication. ITU estimates that 5.3 billion people—66 per cent of the world's population—were using the internet in 2022. This represents an increase of 24 per cent since 2019, with 1.1 billion people estimated to have come online during that period. However, this leaves 2.7 billion people still offline. The effect of connectivity and communication by digital technologies has enabled the human kind with a magical intensity of staying connected.

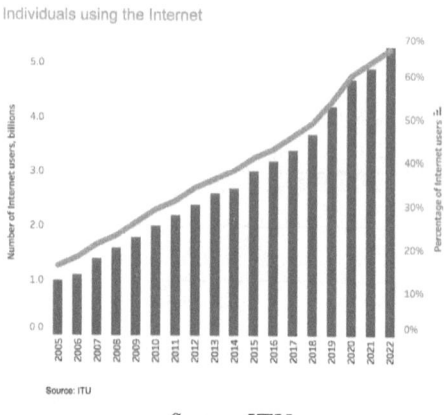

Source: ITU

Portable digital devices. Below is the forecast number of mobile devices worldwide from 2020 to 2025 (in billions) as digital technology puts everything at your fingertips.

Balanced Breakthrough Responsible Technology

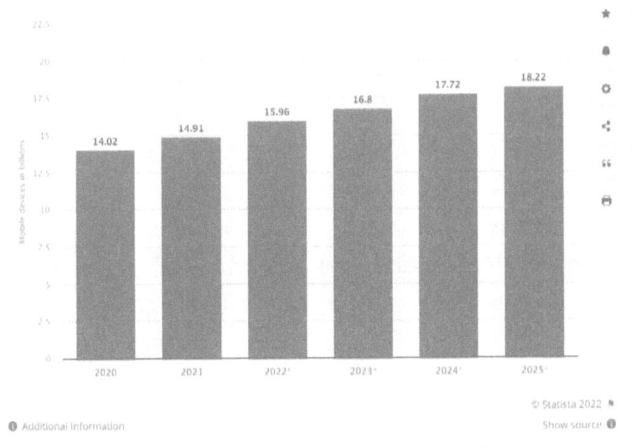

GPS and mapping. GPS has made our environment a safer and easier place to live. GPS is being used to help parents find and keep track of their children and is being installed as a location device in cars and cell phones to assist people in mapping and directions. The most commonly used GPS and maps system is Google Maps. More than a billion people use Google Maps every month, and 5 million active apps and websites use Google Maps Platform core products weekly.

E-books and e-learning. The pandemic has restructured the way students and others learn, and digital method have been helping them to learn without hurdles. Avid book readers have gone through a paradigm shift to electronic devices from paperback books. Studies suggest that the corporate e-learning market could increase by $38.09 billion between 2020 and 2024. The last decade saw the number of ebooks sold per year nearly triple, from 69 million in 2010 to 191 million in 2020.

Entertainment. Digital technology helps pursue hobbies such as singing, playing video games, or creating a live streaming channel from any part of the world. Revenue channels can be created through, for example, YouTube. The amount of money a YouTuber can make per video depends on a variety of factors, such as the number of views they accumulate and how many Google ads that are displayed throughout their videos. At an estimated pay rate of $5 per 1,000 views, a YouTube video with a million views can make upward of $5,000 (Forbes 2018) which makes being a modern-day influencer a pretty lucrative job! (These are only estimates, and some YouTubers may make more or less, depending on the quality of the ad, click-through rate, and other factors.)

Transportation. Several modes of transportation, including airplanes, trains, and ships, use digital technology to accurately navigate routes by sea and land. In the near future, cars and buses are expected to become fully automated, and 55 per cent of small businesses believe that they will have a fully autonomous fleet in the next two decades (source: Nissan). However, only 57 percent of people familiar with self-driving cars would be willing to ride in them, according to a poll cited by *US News & World Report*.

Healthcare. Digital technology lets us do things differently and it has huge potential to help achieve net zero in the Healthcare sector, when used responsibly. An extensive view of patient health is provided by those technologies to the healthcare providers. This significantly increases access to health data for the healthcare providers and giving greater control to the patients over their health. As a result, an increased efficiency and improved medical outcomes are realized.

And there are many other industry segments where digital technologies have been a saviour.

The Rise of Balanced-Breakthrough Responsible Digital Technology

The Sustainable Development Goals or Global Goals (SDG) are seventeen interlinked global goals collectively designed to be a "shared blueprint for peace and prosperity for people and the planet, now and into the future" (source: United Nations). The SDGs were set up in 2015 by the United Nations General Assembly and are intended to be achieved by 2030.

Fig. 3.1 *The seventeen Sustainable Development Goals established by the United Nations for 2030* (courtesy of the Division for Sustainable Development Goals, UNDESA)

Defining Balanced Breakthrough

Together as a society, we need to produce digital products and consume them in a responsible way, treating computational resources as finite and precious. This need should prompt the development of breakthrough solutions toward responsible digital technology.

Frugality and a clean-energy mindset need to be woven into software engineering at all levels so that resources are utilized only when necessary and as efficiently as possible. However, quality should not be compromised—the same results must be achieved, but using less energy.

What I am calling **B**alanced-**B**reakthrough **Re**sponsible **Di**gital **T**echnology, or 2BReDiT (pronounced 'two bred it'), is a rightly blended way of digital technology with an ethical approach to creating, leveraging, or consuming, and distributing new or vintage technology, services, and solutions, with a focus on sustainability and well-being. 2BReDiT also means protecting data; ensuring safety and security as well as respectfulness; adhering to environmental, social, and governance (ESG) standards; and the creation and use of clean energy.

"There are many routes to net zero, but digital technology has a central role to play, no matter what sector or country you look at," asserts Andy Hopper, FREng, FRS, vice president of the Royal Society and professor of computer technology, University of Cambridge. Balanced-Breakthrough Responsible Digital Technology could come in many forms: responsible innovation strategy, responsible project management, responsible product development, responsible frugal computing ... countless actions to ensure the right technology is delivered to the customer and is sustainable.

Ethical AI for Humans and the Environment

The collective ingenuity and the power of diversity truly harnessed in today's fast-paced technological world should help us take a step back and be cognizant of sustainable goals and impacts on human well-being. The empowerment of intelligent systems paves the way for cascading impacts on human life and the environment. Artificial intelligence systems and similar smart technologies should be focused on human-centric collaboration—the trinity of business, technology, and humans—and the democratization of ethical power. AI needs to be trained in ethical decisions along with radical thinking and sustainable impacts.

The computer chips made today have approximately 1 million transistors more than those made in 1970, making them a million times more powerful. This increase in power calls for higher standards for responsible technology in all applications and software—standards that need to be imbibed in every part of the digital product value chain and measured with sustainable key performance indicators (KPIs).

Balance Theory for Responsible Digital Technology

Digital technology includes your mobile phone, tablet, laptop, social media, artificial intelligence—anything digital that can ease your life and add rapid value to your life or work. But for technology responsible to human concerns, the right balance must be achieved in creating this technology, consuming it, and understanding the feedback through intelligent feedback mechanisms. Balance theory can lead to developing a responsible approach.

Let's look at what some of these balance theories involve.

POX Model

Fritz Heider modelled a theory of attitude change in the psychology of motivation. For example, a Person (P) who likes (+) an Other (O) person will be balanced by the same valence attitude on behalf of the other. Symbolically, $P(+) > O$ and $P< (+) O$ results in psychological balance.

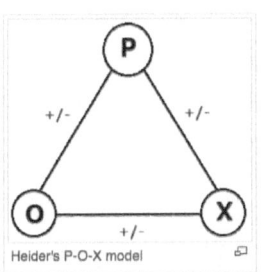

Source: Wikipedia

Balance theory is useful, for example, in examining how celebrity endorsement affects consumers' attitudes toward products. If a person likes a celebrity and perceives that the celebrity likes a product, that person will tend to like the product more in order to achieve psychological balance. However, if the person already had a dislike for the product endorsed by the celebrity, they may begin disliking the celebrity, also to achieve psychological balance. Heider's balance theory can explain why holding the same negative attitudes as others promotes closeness, e.g. "The enemy of my enemy is my friend" (Wikipedia s.v. "Balance theory").

The balance theory has a vast significance in real world and the application of the theory to leverage technology can lead to a responsible usage and a minimalistic approach for consumption of digital technology

Applying Balance Theories

The implications of balance theory for responsible technology could be huge. The proliferation of digital needs course corrections for processing innovation, avoiding overengineering while developing products and trying to build optimized intelligent feedback loops. The key principles should be openness and playing on the strengths of people, technology, applications, and process. Though the pace digital is accelerating, it's important to step back and think over the implications of technology on sustainability and human well-being. Deprioritized business expansion focus could help with open-source business innovation and a greener future in the long run.

The accompanying diagram represents the balance to be maintained for creating and curating digital technologies with the given players of the context—including self, team, and end consumer) across the dimensions of physical well-being, mental well-being, societal well-being, climate change, and sustainability/environment.

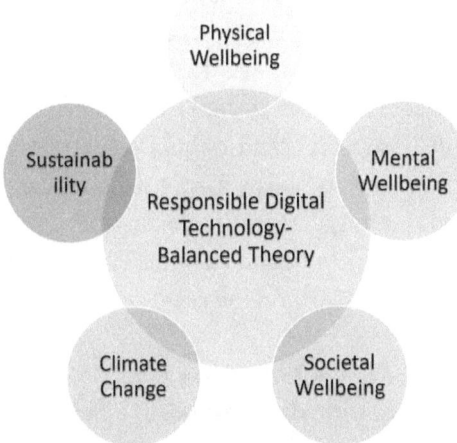

Several balance theories have the capacity to promote an objective of responsible technology with built-in KPIs. These include balanced product development theory, work-life balance theory, frugal computing balance theory, developer coding principles balance theory, user experience balance theory, mobile apps balance theory, legacy application transformation balance theory, and integration as a service balance theory.

Curated Experience and Need for Governance

The detailed balance theories can be extended from here and applied to responsible technologies to create the right mix. AI can be used for acquiring intelligent feedback throughout the process leading to course corrections of the technologies already in operation.

A lot of observations are found by the Technologists and End Consumers around how people are unable to leverage some applications correctly and what kind of drastic impacts that leave. Consider, for example, a real-life use case where a taxi app was having issues finding the location of the end consumer due to the system's incorrect and imprecise mapping. Although it could be just one street away from the location as shown on the map, the confusion related to this behaviour is huge, leading to tensions, cost implications, driver's non-satisfaction, and so on. A small technical issue can leave a drastic effect on the people using the service.

This use case has a lot of course corrections becoming a vital use case for providing responsible digital technology. There could be a possibility of leveraging AI out of the app probably, if not possible from the car taxi application, but like societal governance that will ensure the technology is more responsible than a completely overengineered platform and do the needful for the humankind and environment.

Though there are multiple standards and consortiums in place, it becomes a vital factor for every organization to define a consortium or inherit the right responsibility principles from other consortiums with the objective of sustainable development goals and well-being of humankind including a superior experience for the customer.

The accompanying diagram illustrates the emerging digital technologies that can drive decarbonization and there is a need for making it responsible with near zero goal objective and human well-being:

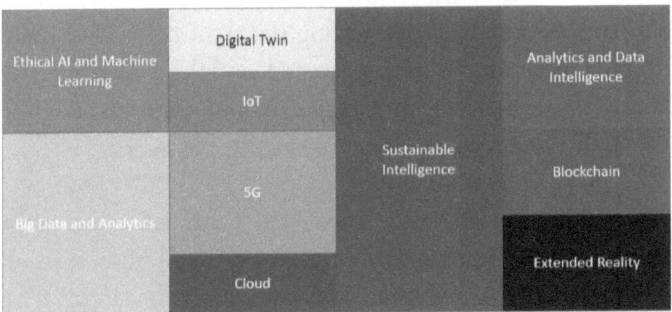

Human Intelligence, Innovative Mindset, and Scientific IQ

Human intelligence is the product of *experience/learning + curiosity + effort* that permits us to pursue objectives in a context. Innovation is very much related to this intelligence, and everything or anything can be an innovation. The inculcation of an innovative mindset through proven techniques like design thinking can help widen human intelligence so that we design machines only as stupid boxes that can supplement human intelligence rather than decrement.

Consider a real-life use case. A 65-year-old man was trying to use an internet banking application. In a form where the debit card details were requested, the first four characters were expected by the app's developer. However, the validations were not in place, and the senior citizen was entering all sixteen digits in that text box. When the senior citizen was submitting the form, the error messages were not properly conveyed. The senior person tried two more times, thinking he had typed in the wrong numbers, and the account got locked. This made him suffer for many days without doing any transactions and not getting the right help. He finally found a millennial whom he trusted, and the millennial was able to understand the issue, keyed in four numbers at a time manually, and recovered the transaction password.

The senior citizen was frustrated with what he went through and expressed his disappointment that a small error impacted him in manifold ways. For a developer, it's a simple validation, but for the end consumer, it's a loss with an impact on his financials and routine life. Although there could have been checks and validations, this error with the payment field type is made to production(live environment). The responsibility principle here would be to think always from the customer's perspective as ensured by automation testing, AI testing, or smoke tests.

The engineering or X factor or product development creation and design principles should be formulated with responsibility at the core. When the scientific IQ of every individual and balanced-breakthrough responsible digital technology are put together on a massive scale, clean energy solutions to consuming and developing can help avoid transmitting greenhouse gases to the atmosphere, control pollution, and promote a green society.

Techniques for Responsible Technology Consumption

There are three main techniques we can develop to balance our technology consumption.

1. Tackle the Culture of Imbalance.

Work-life balance in an imbalanced culture, as noted in the book by Donnie Hutchinson, is one of the key struggles that most employees underwent during the pandemic and in its aftermath. The right usage of technology comes from self-discipline and the will to focus on the task that you are working on.

Alert notifications can easily cause you to deviate focus from your current task. Assign yourselves with windows including Life Windows and Work Windows and ensure the focus does not deviate from one window to the other, a window being a timeslot for work and life in a 24-hour day. However, define the notifications properly so that an emergency call from your kid's school is something that you can give attention to in your work window, and vice versa.

Mobile is the new normal—and more than normal now as we have inculcated the habit of attending to mobile phones into our cognition. Try to eliminate notifications from apps as much as possible or at least prioritize within your life and work windows so that you can respond appropriately.

A healthy balanced diet, at least ten minutes of meditation, at least twenty minutes of exercise, and using technology responsibly can help alleviate stress and allow you to enjoy a positive work-life balance and ambiance.

2. Empower Your Brain.

Too much screen time and doomscrolling have a direct impact on physical and mental well-being. Social media algorithms are intended to make you scroll more, providing too much content that becomes addictive. Diminished brain power and fearful thoughts increase stress. Reducing exposure to digital screens such as phones or tablets before bed reduces the blue light that could damage your eyes. Eating a balanced diet with proteins and other nutrients can help to improve brain power.

The human brain has a memory capacity equal to 2.5 petabytes. The right use of technology can help power up the brain with multiple dimensions of good thoughts so that you can enjoy a happier life.

3. Use artificial intelligence and technology as an aid.

There is a common misconception that artificial intelligence will be a replacement for humans. However, AI should be used only as an aid to humans to ease large-scale manual work by eliminating some strenuous validations. For example, an apparel retailer based in Europe is leveraging AI in AI-driven demand prediction, to optimize supply chain to make sure that they produce the right products for their customers, to the right store, and at the right time.

The 2BReDiT Framework   4

Let's begin this chapter by looking at five goals toward achieving responsible digital technology through a balanced breakthrough approach.

1. **Technology democratization.** Enable responsible digital technology for everyone with an optimized or near zero impact to further sustainable development goals.

2. **Technology balancing.** Curate usage of available technologies within the twenty-four-hour cycle by focusing on the role of fossil fuels, digital consumption, and human well-being. The impact of fossil fuels is everywhere: toothbrushes, buildings made of cement, travel, software computing.

3. **Technology collective ingenuity.** Become aware of the possibility of responsible digital technologies, and cultivate a collective societal conscience that will power the attainment of sustainable goals and human well-being.

4. **Technology innovations.** Use the power of true scientific intelligence as a lever within contextual domains and a sustainable process, with an exponential innovation arm to generate create ideas that are good for the future.

5. **Technology circularity.** Embrace digital technologies for a circular economy and leverage its true life before expanding to another version of the technology.

Building the 2BReDiT Framework

Figure 4.1 illustrates a responsible technology framework based on responsibility balance theory, which can help humans leverage technology in the right balance for the well-being of self, society, and the environment.

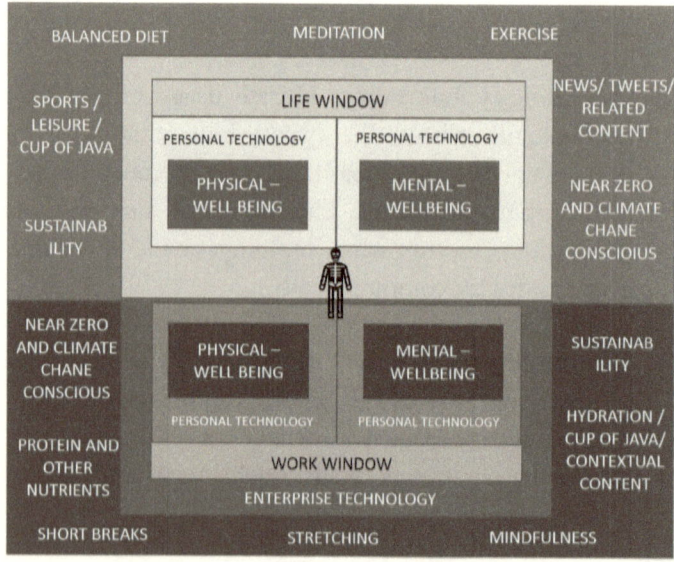

Fig. 4.1 A responsible technology framework based on balance theory

The framework comprises two work-life windows to start with. You can create your own work window and life window for a twenty-four hour period. Here are two windows that might be considered typical.

> **Work window:** eight hours of work, 9 a.m. to 5 p.m.
>
> **Life window:** minimal sleep of eight hours, with the remainder of the time utilized for pursuing other objectives such as travels, 5 p.m. to 9 a.m.

Technology can be leveraged in these windows as appropriate, keeping physical and mental well-being and sustainability in mind. The body and mind together can work in a symphony when balanced well with proper sleep, diet, exercise, and digital detox. Sleep can be improved by eliminating blue light generated by digital devices before bedtime. It's always a good practice to wind down at least an hour before going to bed and to reduce overall screen time as well.

Technology blocks can be used positively within the life window to promote physical and mental well-being, e.g. an app for fitness routines, a YouTube video guiding meditation, a news app for the day's top headlines, and so on. Fitness apps are particularly useful, as they may provide not only exercise or stretching routines but also information, such as an article on the fitness routine promoted by your organization, and goal tracking. The technology can be consumed responsibly by the users and that way you will imbibe a practice of using technology only when needed.

A separate technology block would be tied to the work window, such as your Enterprise Resource Planning application (ERP). In software companies, in particular, the work window is tightly bound with software to develop code, products, platforms, and so on. This use of technology cannot be eliminated, but businesses should consider the employee's well-being and the effects of longer exposure to software routines. On the other hand, some companies, such as in the manufacturing industry, could rely on robotic process automation, as an example, and the dependency on the technology would be handled by the enterprises.

Following are some of the key performance indicators (KPIs) for digital products to measure sustainability by enterprise technology blocks across the value chain:

- carbon footprint
- energy consumption
- product recycling rate
- saving levels resulting from conservation and improvement efforts
- Supplier Environmental Sustainability Index
- supply chain miles
- water footprint
- waste reduction rate
- waste recycling rate

Tracking KPIs is well-known to be crucial for any business. However, many ignore the potential for measuring sustainability KPIs, which allow you to track sustainability progress and potential for improvement in the future.

Examples of Promoting Sustainability

Digital phones and other portable electronic devices can be designed in such a way that, for example, the phone's lifetime is extended, which can lead to a significant reduction of greenhouse gas emissions per year of use across the entire life cycle of the phone. The use of recycled plastics for digital phones can also promote sustainability.

Following are some of the technology companies that Earth.org evaluated in terms of their environmental performance through 2021 (Wong 2022).

- **Apple.** Since 2015, Apple's carbon footprint has decreased by 40%, making steady progress toward its 2030 carbon neutrality goal.

- **Dell.** Dell is ranked one of the top sustainable tech companies for its efforts and is currently partnering with its suppliers to reduce its greenhouse gas emissions by 60% per unit of revenue by 2030 and to achieve net-zero greenhouse gas emissions by 2050.

- **HP.** In April 2021, HP announced its new goals of reaching carbon neutrality in its operations by 2025 and reducing its value chain greenhouse gas emissions by 50% by 2030. HP also strives to achieve net-zero greenhouse gas emissions across its value chain by 2040.

- **Lenovo.** In 2020, Lenovo achieved an overall non-hazardous waste reuse and recycling rate of 88.6%, exceeding its original goal and showing overall success.

- **Microsoft.** Microsoft invested $10 million to support innovative technologies for water conservation, access, and quality while increasing their replenishment project portfolio by nearly 700% from 2019 to nearly twenty replenishment projects.

And many more technological companies are working on greenhouse gas reduction (GHG) projects and realizing their vision.

The Digital Nemesis of Human Well-being

Usage of digital devices has created the modern digital world. It is a fact that with every advancement and technological development, there is an adverse impact on physical and mental health. Digital devices—including mobile phones, tablets, and virtual reality devices—are not an exception. Various studies have highlighted the negative effects of mobile phone exposure on human health, and concerns about possible hazards related to mobile exposure have been growing.

In addition, there are about 2.5 quintillion bytes of data created every day that is unsupervised, which results in content moderation as one of the core propaganda. Studies suggest that content moderators have been affected by post-traumatic stress disorder (PTSD) resulting from exposure to a vast amount of violent content, and many organizations have quit the content moderation business although it's an avenue for exponential business growth. There are potential professional hazards as well.

AI is being looked at as a way to handle the first cut of adverse content before it is made viewable by the content moderators. It will take AI and humans together to solve such problems to ensure the well-being of humankind from mental and physical health perspectives.

Improving Concentration Skills

A number of factors affect concentration. One of these is distraction, when a constant flow of information is bombarding our brains and impairing our ability to concentrate. Others include insufficient sleep, insufficient physical activity, improper eating habits, and environments such as noise levels. Tight focus and clear thinking form the basis for the successful accomplishment of any task or given objective. However, achieving this cognitive state of mind in order to achieve high targets requires putting a number of measures into practice.

The 2BReDiT framework can be of help in determining how you can leverage technology in your work, life, and environmental well-being. Here are nine measures for improving and moving away from negatively impactful habits.

1. **Digital Detox**. Take a break from using electronic devices or certain media for a period of time, whether a few days or several months. The specifics of digital detoxing will differ from person to person, and you

should identify your own ways for how you can be away from digital devices.

2. **Reducing multitasking.** Multitasking—the attempting to perform multiple activities at the same time—makes us *feel* productive. However, it's also a recipe for less focus, poor concentration, and lower productivity. True multitasking is achieved by performing multiple tasks efficiently by focusing on each one singly within a given time frame. The goal is high productivity levels and high-quality outputs, not doing everything at once.

3. **Living the moment.** Choosing to focus on the current moment is one of the greatest "secret sauces" to help you to achieve and go places. It's tough to concentrate when your mind is always in the past and worrying about the future. Let go of things sometimes and live in the current moment with the concentration to accomplish your goals.

4. **Physical health and brain-training exercises.** Listening to music, solving a Rubik's cube, and eating a well-balanced diet can help improve focus and concentration in manifold ways. Scientific research is starting to amass evidence on the ability of brain training activities to enhance cognitive abilities. Puzzles such as a Rubik's cube can help train your brain to hyper-focus on a task. Using a timer can help you to be more time-focused and complete the work on time.

Balanced Breakthrough Responsible Technology

5. **Eliminating distractions.** Configure alerts and notifications properly in your digital devices with the right priorities.

6. **Mindfulness and meditation.** Every day, devote at least ten to fifteen minutes to meditation and mindfulness. A strict daily practice can help bring a good change within you.

7. **Good sleep.** Do not use digital devices before bedtime, as the blue light generated from these devices may harm the eyes. It can create disruption to the quotidian cycle that affects sleep routines, as blue light mimics the light of day. It would be good instead to read books of a soothing genre, listen to music, even if using Alexa, or meditate. It is good to have eight hours of undisturbed sleep, so develop a good bedtime routine.

8. **Short breaks.** Researchers have found that our brains tend to ignore sources of constant stimulation. Taking very small breaks by refocusing your attention elsewhere can dramatically improve mental concentration when you return to the task.

9. **Connecting with nature.** Finding time to take a walk in the park or appreciating the plants or flowers in your garden can boost your concentration and help you feel refreshed. Research shows that just having plants in office spaces can help increase concentration and productivity, as well as workplace satisfaction and better indoor air quality.

Principles of Frugal Design

5

Technology companies are at a critical juncture which demands future-proofing by CXOs and IT leaders to ensure adaptability and resilience. However, future-proofing is the eternal conundrum of digital technology. Multiple organizations and governments have been asking how digital technology can solve the puzzle of promoting sustainability and human well-being. The world is reaching a saturation point, and the entire world is focused on reducing global warming. Casting digital technology as a saviour is a good start toward delivering outcomes and promoting sustainability.

Leaders need to unlock hidden opportunities and redirect their enterprises to counter extreme disruptions and grow stronger. Adaptation through modularity can be a force multiplier for the leaders of an organization to exploit the disruption and convert it into potential opportunities. Think of the components of modularity as Lego blocks, empowering your network with collective ingenuity and discovering how to value the customers and employees, with sustainability at its core.

Responsible Enterprise Technology through Frugality

In a modular approach, the enterprise technology block holds a major responsibility for developing software, products, platforms, mobile apps, and so on. The goals of enterprises have changed, and many enterprises are adopting sustainable initiatives and reducing their carbon footprint. Being economical with the software leverage and optimal resource consumption by software defines the frugality that needs to be woven into software engineering. In simple terms, frugality can be sustainable initiatives and Green Software state that can be devised with Green Software principles while developing and maintaining software.

Enterprises of all sizes are in the process of retiring or transforming from vintage monolith platforms or applications to modern architecture patterns of a single or distributed cloud and edge computing. On the other hand, the "everything as a service" model (XaaS) has paved the way to greater business potential, bringing services as close as possible to a digital reality and innovating business models. However, continual digital demand for production leads to increased consumption of data and computing resources. The technological debts with bimodal architecture have intensified this demand and remain still a challenge. Achieving a near-zero tech debt is one of the prime objectives for the digital tech-savvy or non-digital natives.

Responsible technology imposes a set of considerations for reducing technological debts and moving towards sustainable computing. Following are some of the architecture and design considerations based on the responsible technology balance theory.

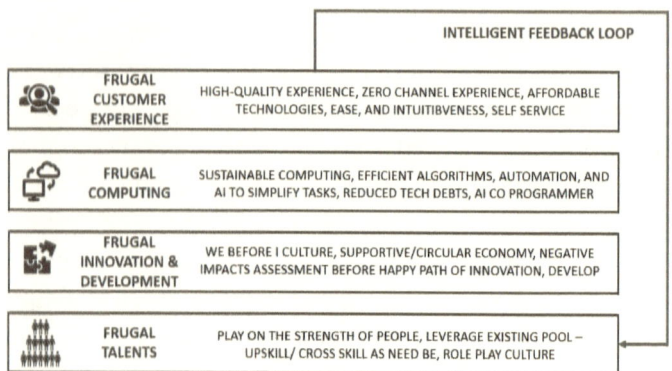

Frugal Talent

Frugal talents are highly responsible and don't let technology interfere with them when focused on work or life routines. When using digital devices, they try to keep their focus on the content of the appropriate window by switching off or managing alert notifications. They may have their own habits of resuming technology and redirecting their attention when needed.

Following are sixteen principles for leveraging the existing talent pool within your organization.

1. Play on the best strengths of the people in your organization. They all have their strengths.
2. Know the limits and boundaries of the human context you are dealing with.

Balanced Breakthrough Responsible Technology

3. Be innovative and deliver the right value. Accept people in their current form, and try to enhance their skills and knowledge if they want.
4. Listening is a key skill. Acquire the broader context before you judge.
5. Try to avoid passing the wrong messages.
6. Pass on your knowledge to others. Remember the saying: to teach someone to grow or make food in the long run is better than feeding them right now.
7. Embrace questions and conditionals: what, why, if, when, and so on.
8. Failure is temporary, but if you give up, you make it permanent.
9. Ensure the mental and physical well-being of everyone.
10. Don't try to do everything by yourself. Collective efforts could create magic.
11. If you have ten hours to cut a tree, take seven hours to sharpen the axe, and you will cut down it in three hours. But if you spend eight or nine hours on the sharpening, you risk not being able to finish your task.
12. Provide cross-skill and upskill opportunities as needed.
13. Try to build an army of your peers from day one. Slow, steady, and progressive could bring wonders.
14. Practise "We before I," and stand together as "we."
15. Don't worry about the future too much. Learn from the past and build the future, but live in your current situation.
16. Respect others' innovation, IP, cause, and morals, leveraging them within their legal rights.

Frugal Innovation and Development

A collaborative team culture with an innovative mindset helps to mix and match the capabilities, functions, tools, and processes for developing and delivering the intended software. The reduction of technology debts and efficient use of resources, innovative thinking, and collaborative team effort will help to innovate and develop software from a perspective of sustainability that could form a core for future development methodologies.

The innovations need to be approached with complete awareness of their implications. Innovation could be anything, from the colour of a simple button to an e-delivery app. However, any negative implications should be thought through thoroughly and tackled while the creative path flows happily.

Let's look at an example. A social singing app might be a perfect business innovation in support of an existing hobby that potential users already pursue. But possible impacts might be overlooked, such as an addiction to the singing app that would be an indirect reflection on the husband-and-wife relationship. The human brain generates dopamine when there is happiness, so even too much singing can become an addiction. The principles of responsibility should play together in the form of intelligent feedback loops focused on the well-being of the end consumers and not just on earning more money.

The core responsible philosophy of frugal innovation and development lies in vetting for end user-focused responsible innovation, plagiarism, and respecting intellectual property and copyrights. Here are fifteen further principles of frugal innovation and development.

1. Be aware of the carbon footprint of the innovation or solution in moving towards net zero.
2. Try to avoid conflicts and respect everybody's opinion while aiming for the right decision.
3. Put your independent thinking first, but believe in collective effort and brainstorming before you retreat into silos.
4. Always practise role play and design thinking.
5. Mistakes will be made, no matter how much care is taken to avoid them. But remember to learn from a mistake and try not to repeat it. L *Don't react disproportionately* or *Limit the volatility of environments in which mistakes are detected and responded to.*
6. Innovation can come in any form. Adapt, take a cautious decision on the value it generates, and mature.
7. Always look out for generics and try to create instances out of it. Repeat if it works well.
8. Try to grow business and the domain knowledge which is an essential core.
9. Keep it simple. Communicate with your audience in a way they can understand.
10. Try to leverage best practices in any form, and apply them with a conscience for the relevant context.
11. Don't point fingers. Focus on how *we* can handle the current situation in a better way.
12. Neither you nor I will always be right, but together we need to ensure the right value to the end user.

13. Have your disclaimers in the right form, set hopes at the right level, and paint the pictures in the right colour.
14. Try to make elevated pitches and detail-oriented pitches in the right context.
15. Manage and care for devices and data.

Frugal Computing

The principle of frugality woven into software engineering allows for sustainability in the long run, and it's an emerging discipline. Following are ten possible actions for developing methodologies and the value chain.

1. Eliminate cargo cult technology and programming that resides as legacy code but is never used.
2. In e-commerce, use robust modelling and AI-driven prediction of customer behaviour to make shoppers aware of how they can be a part of big-picture eco-friendly sustainability initiatives.
3. Use AI pair programming to collect feedback from prod incidents and for quality assurance, and to suggest code and entire functions in real time.
4. Reimagine business processes leveraging the cloud where the cloud is not looked at as a data centre.
5. Define, build, and run sustainable software applications ("The Principles of Sustainable Software Engineering" from Microsoft is a good place to start with learning how).
6. Weave reusable components for repeated operations into coding practice for eliminating field mappings or settings (abstraction).

7. Refine power-intensive algorithms and processes for software platforms through optimizations on the model of Proof of Work (PoW) to Proof of Stake (PoS) consensus for Ethereum for digital platforms such as blockchain.
8. Embed carbon emission intelligence in your applications and software
9. Build robust real-time carbon footprint tracking intelligence in transport and logistics, and act to reduce emissions with an optimized path, mode, or solution.
10. Acquire an essential understanding of carbon emissions and carbon footprint from software applications, supply chains and plants or facilities, including electricity usage, heating, cooling, and ventilation, towards enacting clean energy measures.

Here are four broad considerations to look at in frugal computing:

- automation or AI as an aid for technology ease
- sustainable computing and use of efficient algorithms
- clear use of data and appropriate geo-policy storage and backup
- collateral damages versus silver linings

Frugal Customer Experience

The end customer experience should not be compromised with less usage of resources. The frugal customer experience should always be high quality, accessible from any channel, easily used, and enabled towards self-service with optimized resource usage. Responsible technology principles should include the well-being of the end customer and the impact left on the society and environment (ESG).

Intelligent feedback loops through artificial intelligence should capture how the user is experiencing the developed software or application, monitor robustly in the background with respect to the sensitivity of the data, and provide the right help as needed. Create new stories to tell and inculcate storytelling. Always put yourself in the customer's shoes. Imagine that the customer is you, and design for yourself.

Here are three aspects of frugal customer experience to keep in mind:

- continuous feedback and frequent self-assessment, preferably AI-driven
- high-quality and any-channel experience
- easy, intuitive, and self-service–enabled experience

Frugality: A Principle of Responsible Technology

As today's digital world accelerates at a strong pace, it becomes imperative that we pass on the right technology values to the younger generation. The generational cohorts known as the baby boomers, Gen X, and millennials have seen a world without smartphones and a world with smartphones, the metaverse, and so on. However, most Gen Z and the Alpha Generation have not seen a world without smartphones, and digital channels in some form are inherent to their lives from the start. It is a great responsibility of the preceding generations to help the new by passing on the right tech values and ensuring the responsible use of technology.

The world is in need of sustainable computing and has been already marching towards the path of zero-carbon computing. The principle of frugality as outlined in this chapter can be a guiding mechanism for inculcating a collaborative culture and designing systems for producing as well as consuming technology responsibly.

Setting an objective of sustainability for digital production as measured by key performance indicators would be a real game changer. The KPIs today associated with digital products are the velocity of features, deployment times, and request times, among others. However, there is a dire need to have more sustainability-related KPIs, including carbon footprint, dashboards of climate science, usage of electricity markets, and data centre design. These would help to inculcate a sense of responsibility within every software engineer, digital domains lead, and the entire organization to create standards and forums for producing technology responsibly and course-correcting for any deviations, as need be.

Green Consumerism as a Business Model

A carbon offset is a reduction or removal of emissions of carbon dioxide or other greenhouse gases made in order to compensate for emissions made elsewhere. Offsets are measured in tonnes of dioxide equivalent (CO_2e). One tonne of carbon offset represents the reduction or removal of one ton of carbon dioxide or its equivalent in other greenhouse gases (Wikipedia 2022).

Carbon offset is essential for achieving net zero goals, and multiple initiatives have been greenlighted by countries and organizations around the world. It then becomes the responsibility of every individual for carbon offset to become carbon neutral.

A Green Consumerism Business Model

In the D-VUCAD (disruption, volatility, uncertainty, complexity, ambiguity, and diversity) world, supply chains are volatile and susceptible to disruptions. However, the e-commerce shift is one of the keys to resilience and

Balanced Breakthrough Responsible Technology

effective supply chain management. E-commerce also helps to promote green consumerism to a larger extent.

The business model in a carbon-neutral world demands end-to-end processes and the promotion of sustainability at the core. The accompanying diagram provides a potential business model in which green consumerism can be promoted to end consumers along with carbon offsets and a new avenue for business growth.

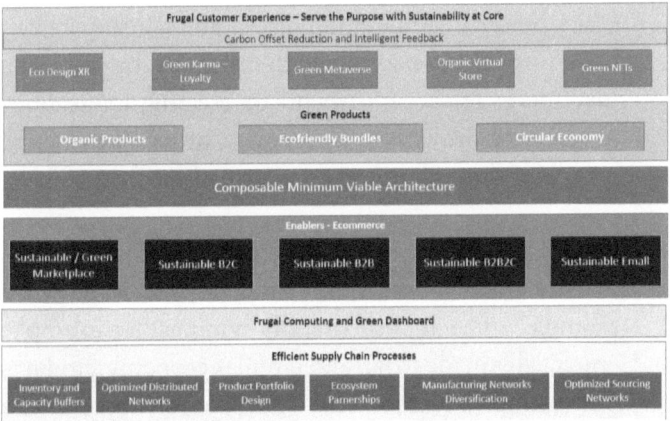

A frugal customer experience lies at the core of this model wherein customers experience what they need at a high level of quality and incidentally contribute to the planet. Achieving this frugality requires a multilayered approach comprising multiple layers including Expereince, Green Products, Architecture, E-Commerce, Computing and Supplychain management.

1. **Experience.** This layer focuses on providing the best customer experience with a green mindset, including Eco Design XR, Green Karma loyalty rewards, Green

Metaverse, Organic Virtual Store, Green NFTs, and others that help to nudge the customers to experience green products.

The experience layer also involves identifying green product usage using AI feedback loops; calculating carbon footprints such as electricity bills and travel charges; and identifying the carbon offset.

Green Karma loyalty points can be awarded for using green products and carbon offset, and consumers can be rewarded, for example with offers on the products. Just Energy offers a framework for creating applications to calculate the carbon footprint at https://justenergy.com/blog/how-to-calculate-your-carbon-footprint/.

2. **Green products.** This layer hosts all the products that are green in nature, including organic produce (for example, spinach), eco-friendly bundles (for example, organically made toys), and circular economy products (for example, furniture as a service, or apparel as a service).

3. **Architecture.** This layer hosts the entire business model in a composable minimum viable architecture for doing composable business by mixing and matching business functions using microservices or serverless functions on the cloud.

4. **E-commerce.** The global e-commerce market in 2021 was valued at USD 14.30 trillion, and is likely to reach USD 52.06 trillion by 2027 (source: GlobeNewswire). E-commerce has proven its merits as a platform for

buying and selling. This layer onboards multiple business models within e-commerce, including Marketplace, B2C, B2B, B2B2C, and Emall.

5. **Computing.** The frugal computing layer is where carbon emissions from computing are minimized by means of effective cloud computing techniques and optimized algorithms. In addition, green dashboards are constructed to ensure the carbon offset is achieved.

6. **Supply chain management.** Efficient supply chain management is needed for handling disruptions and achieving resilience in changing times. This layer includes inventory and capacity buffers, manufacturing network diversification, optimized sourcing networks, optimized distribution networks, product portfolio design, and ecosystem partnerships.

What is digital green karma?

Digital green karma is the benefit received towards human well-being for good deeds in becoming carbon neutral by using digital technology. The good deeds might include saving electricity, consuming natural foods and products, and traveling efficiently without causing damage to the atmosphere.

What is the green premium?

"Green premium" is a term used by Bill Gates in *How to Avoid a Climate Disaster*, referring to additional costs of implementing solutions that are more climate-friendly than their fossil-fuel counterparts.

The approach outlined here focuses on implementing an effective means of reducing green premiums by leveraging existing digital technology solutions while focusing on net-zero objectives.

How do I calculate my carbon footprint?

Alexandra Shimo-Barry, author of *The Environment Equation*, has come up with a formula for calculating your carbon footprint at home.

1. Multiply your monthly electric bill by 105
2. Multiply your monthly gas bill by 105
3. Multiply your monthly oil bill by 113
4. Multiply your total yearly mileage on your car by .79
5. Multiply the number of flights you've taken in the past year (4 hours or less) by 1,100
6. Multiply the number of flights you've taken in the past year (4 hours or more) by 4,400
7. Add 184 if you do NOT recycle newspaper
8. Add 166 if you do NOT recycle aluminum and tin
9. Add 1-8 together for your total carbon footprint

Keep in mind that an "ideal" carbon footprint (or a "low" footprint) is anywhere from 6,000 to 15,999 pounds per year; 16,000–22,000 is considered average. Under 6,000 is considered very low. A footprint over 22,000 calls for some living green practices to be brought into consideration

Becoming Green Warriors

Nations need the military to protect themselves against "bad actors", and the business model proposed above is in correlation with that thought process. The world needs a new military force—the digital green warriors to wage a campaign towards carbon neutral, in the super app context and extending to other carbon-neutral initiatives.

The Digital Green Warriors SuperApp would allow users to calculate their carbon footprint from electricity, travel, and recycling using the environment equation. This data can be inputted by leveraging the end user's bills or by manual input. It can also be retrieved from electricity boards or airports in accordance with the country's regulations and privacy laws. Based on the calculated consumption, the end consumer can be awarded green karma loyalty points.

Digital Green Warriors SuperApp can be a massive green consumerism loyalty program. The super app, being one of the key emerging technologies in 2022 can be pivotal to a frugal customer experience. It provides endless possibilities of onboarding multiple enterprises as a "unicorn" of green consumerism. Green Warriors is just one potential application for creating a green economy. There are thousands of such initiatives being embraced.

A super app could onboard industries that are working in parallel to achieve sustainability objectives, such as the following.

> **Apparel retailers:** 10 per cent off on sustainable fashion; meeting sustainable objectives of enterprises
>
> **Plant delivery services:** plants that improve indoor air quality; eco-friendly delivery
>
> **Organic food services:** awareness of how organic food is healthy and how people can buy local food and support farmers
>
> **Organic beauty products:** natural, organic, and zero-waste make-up and beauty products
>
> **Eco-friendly kids' toys:** toys made of sustainable materials
>
> **Other businesses,** such as an ink refill business that reduces the number of ink containers in landfills by reusing old ink cartridges; sustainable event planning; and food delivery services offering organic and healthy foods

AN APPAREL WORKFLOW – SUPERAPP IN ACTION – CONSUMER JOURNEY OF ALEX

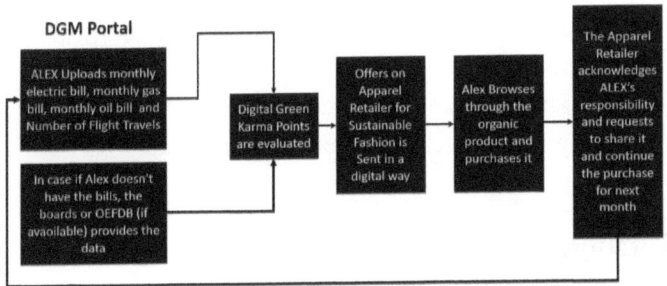

An apparel workflow super app in action:
the potential consumer journey of Alex

DIGITAL GREEN WARRIOR SUPERAPP

The super app awards Green Karma points based on the usage of energy resources. The offers and discounts allocated to the end consumers provide four benefits:

1. **green consumerism**, driving end consumers towards green products;
2. **reduction of greenhouse gases emission** by encouraging a propensity towards saving power with larger benefits and earning Green Karma points for optimized purchasing;
3. **saving the environment** by being part of the social climate to change activities inherently;
4. **promoting sustainability** through green consumerism as a social activity.

The super app benefits enterprises by promoting

- the enterprises' sustainability objectives through participating in reducing their carbon footprint;
- direct business growth in organic product sales;
- participation green consumerism of consumers involved in climate change mitigation.

Intelligent Dashboards

Adaptive AI can be leveraged for building dashboards for both end-consumer perspectives and strategic perspectives to provide abundant insight into how Green Karma points are built every month, leveraged, and redeemed. The analytics could provide details on how these insights can be fed back to marketing to channel more green consumption and reduction of the carbon footprint.

Revenue and Funding

The super app can be developed as an open-source initiative with relevant licensing in one or multiple types of open-source business models, including open core or dual licensing, or a hosted version of the product or donations.

Funding can come from ventures investing in carbon neutral, crowdfunding, enterprise funding, or national or global funding.

Overall, the super app model could be one of the most cost-effective business models for reducing the green premium, enabling a No Poverty movement wherein local farmers have a lot of potential say in producing organic products, reducing carbon footprint, and achieving net-zero goals.

Given the growth of the e-commerce market, every enterprise, nation, and consortium can come forward to fund this model and reduce the green premium and by implementing carbon offset as a strategic objective for net zero.

In summary, the super app:

- helps to leverage technology in the right and responsible way;
- will be the biggest and most massive loyalty program that will extend beyond borders;
- help create a green "military" to save us from the carbon enemy;
- enable growth in the economy and help people fight against poverty in a digital way and for the betterment of lives;
- act as an effective green premium reduction Initiative.

Digital Greening from the Laws of Nature

What are the elements and laws of nature? Almost every ancient culture described natural phenomena as composed of several basic elements. In China, for example, traditional Chinese medicine and feng shui derive from a cultural conception of nature. In Indian philosophy, as reflected in yoga and Ayurveda, the five elements are known as the *pancha bhuta*. These five basic elements are earth, water, fire, air, and space or "ether". They represent the physical and energetic qualities of the human body and of the physical world. They are answers to the question, "How is nature designed?"

The Sanskrit Elements

The five elements of nature are known in Sanskrit as the pancha bhuta or *panchamahabhuta*. They form the basic building blocks of the universe. Every person, animal, plant, and thing is composed of the pancha bhuta in various combinations. Each element has its own characteristics and properties.

- **Earth** (*prithvi* or *bhumi*): solidity, stability, and grounding
- **Water** (*apas* or *jal*): fluidity, adaptability, and change
- **Fire** (*tejas* or *agni*): energy, passion, and transformation
- **Air** (*vayu*): movement, expansion, and communication
- **Space** or **ether** (*akasha*): emptiness, consciousness, and intuition

Every body, whether it is the individual human body or the larger cosmic body, essentially is made of five elements or pancha bhuta—earth, water, fire, air, and *akasha*.

The States of Matter

Each of the five elements represents a state of matter. Earth is not just soil but everything in nature that is solid. Water is everything that is liquid. Air is everything that is a gas.

Fire is that part of nature that transforms one state of matter into another. For example, fire transforms the solid state of water (ice) into liquid water and then into its gaseous state (steam). Withdrawing fire recreates the solid state. Fire is worshipped in many yogic and tantric rituals because it is the means by which we can purify, empower, and control the other states of matter.

Relationships among the Elements

Each of the five elements has a certain relationship with the other elements based on their nature. These relationships form the laws of nature. Some elements are enemies in that one blocks the expression of the other. Fire and water, for example, will "destroy" each other if they get the chance. In order to coexist, fire and water must be separated. Too much fire in the body will create inflammation, while too much water can dampen fire and cause indigestion.

Some elements are said to "love" each other in that they are supportive and nurturing of each other. Earth and water love to "hug" each other, and air and fire increase each other.

Other elements are simply friendly and cooperative. For example, water and air can live together without problems, as they do in soda water; but when the chance occurs, they separate. The same happens with fire and earth.

Space is the mother of the other elements. The experience of space as luminous emptiness is the basis of higher spiritual experiences.

What is the cooperation of the elements of nature in a human body?

These five elements form the world we live in and the structure of our body-mind. If these five elements do not cooperate, one can struggle as much as one wants, and nothing happens. Only with their cooperation, from the basic aspects to the highest aspect, does one's life become a possibility.

Four of the elements form the human body in these proportions:

    72% water

    12% earth

    6% air

    4% fire

The remaining is akasha. You do not have to bother about akasha unless you want to explore mystical dimensions of existence.

Among the four elements, air is the easiest thing to manage because of breath, which you can take charge of in a certain way. Mastery over fire can do many things to you, but because you are householders living in a family situation, you do not have to take charge of fire. To live well, four elements are enough. The fifth element is not relevant for people who doesn't want to explore the mystical dimensions of the existence

How does this cooperation apply to a massive scale when it comes to earth?

Earth, water, and fire are tangible things that can be touched or seen; they exist as matter. Space and air are intangible, yet they exist everywhere around us, even though we cannot see them. Earth, water, and fire are therefore easier for us to understand than space and air because they have more concrete forms. However, all five elements are equally important and interrelated.

The planet Earth is often compared to a majestic blue marble. Water makes up about 71% of the Earth's surface, while the other 29% consists of continents and islands. Earth is surrounded by an atmosphere which brings the other elements into the picture.

What happens when one of the elements is out of synch?

The imbalance of one or more elements in a human body causes problems in one's life. By balancing these elements, the problems can be resolved.

By analogy, the earth's imbalance that we see today lies in the out-of-synch human-created disasters in the atmosphere and the "carbon enemies". The cosmic design by Mother Nature will try to balance on its own by creating havoc unless every individual acts towards green consumerism and an inclusive green economy.

As an analogy, if our kid has a high temperature, will we not act to give medicine? Our Mother Earth has a fever, with an increased temperature—currently of 1° and possibly reaching 1.5°. It's more than time for all of us take responsibility for our own organic green consumerism. Let's make sure Mother Nature's temperature does not get any higher!

"Earth Money" for an Inclusive Green Economy

When I take my daughter Tharika to the park, she plays a restaurant game with me. To her, money is just leaves and nuts and she would trade them as food and cash for paying bills. And I think that is the "earth money" we should all invent, where it could be Organic Vegetables, Organic Clothes, Organic Beauty Products and any organic element that will bring betterment to the well-being and improve the economy.

It is interesting to consider that money has no intrinsic value. Instead, money is an object that has a value placed on it, which allows for the trade of goods and services. Some money, such as metal coins, have actual value in terms of the materials used. However, paper money is more common in the modern world and typically has no real value. Currency has taken several different forms as money has evolved over time.

We are now in a world where paper money is seeing its last stages as technologies like cryptocurrencies have started taking various forms. However, as money becomes more digital, the implications for the green economy seem to be diverging into a negative path. We can anticipate people not living their lives in full, and Generations Z and Alpha experiencing a higher rate of premature death. We must keep in mind that nature's design always allows course corrections, and it's time for every individual to act as we need to. As an outcome, Green Consumerism can be a new way of improving economy and health and wellness.

Balakarthik Baskaran

SLAMIX for the Green Car of the Future

Another possible application of the laws of nature should be in a car design. The design of a green car can be summed up by the acronym SLAMIX.

> **S**un, as a source of energy
> **L**and, for travel on Earth
> **A**ir, to fly in the sky
> **Mix**ture, Mix of Sun, Land and Air modes

SLAMIX is a way to focus on car design based on the principles of nature. SLAMIX could lead to a next-generation car using responsible technology while at the same time taking care of the three Ps: people, profit, and the planet.

Renewable energy sources could lead to a greater green premium. The internals could be based on nature's principles for using the fuels created by the nature—e.g. green hydrogen is created by electrolysis based on water, which is stored and transported. Imagine membranes that create green hydrogen when your car is deeply immersed and traveling in seawater. Research reported by the National Science Foundation (2020) recommends renewable hydrogen fuel from the sea for avoiding the corrosion of electrodes. The possibilities enabled by the core philosophy are limited only to the imagination of the inventors. The land mode could operate with a charging station or electric fuel cars. The air mode could be operated by solar panels. This enables SLAMIX to drive on land, fly in air, and travel under the sea as a green car of the future.

Here I should acknowledge that the starting point of these thoughts came from my son, Nikhil, a Gen Z evangelist for the green car revolution, with some internals worked out by me. I was astonished to see how our next generation is gearing up their thoughts; being a millennial, I have seen a world without smartphones or even a mobile phone, so the ideas of Gen Z and their Alpha successors are beyond my imagination. And yet it has become an immense priority and responsibility for us all to create a world and economy that is greener.

Embracing Change

Change is already happening, and we have to embrace the change. The transition is already happening, and we have to make the transition with the right awareness. Digital "warriors" with phones as swords and green consumerism as their aegis can embrace that change more rapidly.

We can see countries that once rated highly in the happiness index are slipping, and people are becoming more stressed. Amidst the confusion, many people are not coping with their jobs. Baby boomers and millennials are struggling to keep up with the programming skills needed with the advent of new technologies every day. There are a lot of resignations resulting from stress and anxiety, and people are worried about money, which itself could be replaced soon—remember, it does not have intrinsic value. But trends are set, and then they are broken. So technology considered new today will be outpaced very rapidly.

The rapid advancement of technologies will bring new inventions like neural laces, which will make a thing happen when people think it in their mind. These technologies are invented to help differently-abled people. But today's technologies have taken a different shape; many people are becoming differently abled in multiple forms deviating from the original philosophy of technology invention.

In tandem with a popular saying and drawing parallels, I would say, "One person's green consumerism is another life to live." And to quote the great Tamil poet Avvaiyar, from her popular collection of single-line verses *Athichudi*:

இயல்பு அலாதன செய்யேல்

Don't be artificial; be natural.

In terms of digital responsibility, technology should be an aid in supplementing human intelligence, not a detriment.

The Missing Element

The Earth's spheres are classified into 4 subsystems – "lithosphere(land)", "hydrosphere(water)", "biosphere(living organisms)" and "atmosphere (air)". I have strong feeling that the 4 per cent of air or atmosphere that make up the body and nature as a whole relating to the problem of carbon neutral emissions could possibly be solved by using its opposing force, which is fire.

Researchers at Northern Arizona University found that "prescribed burning" can actually reduce the carbon footprint (NAU Review 2010). So there is some evidence of this theory. My point is that if we take a pause and look outside the box at the problem, fire in small, controllable forms can destroy carbon.

Fire is present in multiple forms in today's world, e.g. dynamite, bombs, wildfires, and so on. In my own ancient culture, we used to place *diya* lamps in a diamond-shaped ring arrangement for worshipping god. A scientific way of looking at it might be that the burning of the lamps reduce the negative impacts of carbon. Ancient rituals and festivals from many cultures around the world celebrate fire. As a potential solution to destroy the "carbon enemies", which were once celebrated as diamonds and coals but have taken shape in a different form, we could light candles, diyas, firecrackers, or other flammable devices in the atmosphere or outdoors. The catch here is how to do it controllably and with what frequency, and how to address within every culture the noise levels or the need for pollution-free incendiary devices. Digital technology surely can help do it through such means as predictive learning, adaptive AI, sustainable intelligence, and satellite monitoring.

If green consumerism has to start from the self, then the destruction of the carbon enemy also has to start with the self—but only on a massive scale in the right quantity can it make a difference. That doesn't mean we have to set fire to forests or houses or burn down anything on a large scale. In Indian ancient culture, as expressed in the *Mahabaharata*, the Pandava lit diyas to have a controlled and peaceful ambience. Similar practices are found in Chinese tradition and in other cultures and religions. But this act is not to be perceived as belonging to any religion, caste, nation, or country; rather, it is a means of creating a green world and green economy.

If technology has been the destroyer, creating so many diseases, let technology be the solution.

The world has seen so much destruction with the pandemic, and so many people crying. The persistence of each and every one today should be for a reason. Nature's laws can always govern us with the right means and balance. This balance and the design of nature can't be opposed, but they must be understood if we are to save ourselves and create a greener future. I believe we can live happily with our dears while not missing any of the great inventions that we have today.

This point of view can be supported by technological research among communities within all the generational cohorts—while maintaining their borders and current legacy forms and without creating too many crises and chaos. Any transition from a legacy world is a progressive change. A coexistence model needs to be delivered.

However, combining progressiveness and the adoption of all parts of our ancient cultures can bring lot of synergies and light the true light of the green world. The economics of oils, candles, and organic products can be manifold while bringing smiles and joys as the true deliverables of a breakthrough-technology innovation.

A responsible technology council and communication can be supported by funding from nations coming together along with ancient cultures. Modern technology worlds have seen single-click deployments of software. There is no single click to bring about the green future in one day—but it *will* happen one day.

Being a millennial, I am responsible for creating a new and greater future for the Alpha Generation, my own son and daughter and all their friends in the world, and for protecting the baby boomers, Gen X, and Gen Z with the help of ancient cultures. This "I" should become "we", "you", and "they".

Allow me to share my daughter's favourite song, called "I See the Light":

> All those days watching from the windows
> All those years outside looking in
> All that time never even knowing
> Just how blind I've been
> Now I'm here blinking in the starlight
> Now I'm here suddenly I see
> Standing here it's all so clear
> I'm where I'm meant to be …

And Sudha Murty's words:

> GOOD RELATIONSHIPS, COMPASSION AND PEACE OF MIND ARE MUCH MORE IMPORTANT THAN ACHIEVEMENTS, AWARDS, DEGREES OR MONEY.

It is my hope that this should become a reality one day within all people and by all the people in the world.

Responsible Technology: Forerunners and Present Action

"Responsible AI is better AI." This is how Linda Leopold, the head of Responsible AI & Data at the leading apparel retailer H&M, asks us to think about artificial intelligence: as a toddler with superpowers, "a compelling technology that can have an incredibly positive impact on business and society but is also immature; it needs parenting" (H&M Group 2021). In this chapter, we will be looking at tools and resources for ethical and responsible technology.

Following are some of the forums, responsible tech societies, and sustainability champions across the globe that practice responsible technology, ethical AI, and sustainability goals.

The Swedish government's **Committee for Technological Innovation and Ethics (Komet)** offers a tool in English designed to help you take a responsible technology approach to your work (responsibletech.se). Horizontally, Komet addresses governance innovation in four dimensions: responsible technological development, collaborative public governance, regulatory development, and testing facilitation.

On the vertical axis, Komet focuses on cross-sectorial issues and policy-developing initiatives in which new technology contributes to digital transformation, climate change, and health.

Supporting industrial AI for better and more secure societies is the focus of **Qamcom – Unity of Technology Experts**. "Qamcom offers intelligent methods for creating more efficient processes and smarter solutions to those who understand how and when AI should be implemented"

The mission of the **Institute for Responsible Technology** (IRT) is to protect genetic integrity and nature's biological evolution by preventing the outdoor release of genetically modified organisms, and to protect human and animal health by preventing the use of GMOs in the food and feed supply (responsibletechnology.org).

CodeCarbon is described as "a lightweight software package that seamlessly integrates into your Python codebase" and "estimates the amount of carbon dioxide (CO_2) produced by the cloud or personal computing resources used to execute the code" (codecarbon.io).

Climatiq "provides embedded carbon intelligence software that enables developers to automate GHG emission calculations based on verified scientific models. Its suite of products includes an open-source dataset of emission factors and intelligent APIs that integrate with any existing software for real-time monitoring of greenhouse gas emissions" (www.climatiq.io).

"The **Google carbon footprint dashboard** displays estimated greenhouse gas emissions associated with using Google Cloud services for your account (cloud.google.com/carbon-footprint)."

"The **Emissions Impact Dashboard of Microsoft** also helps calculate your cloud-based carbon emissions (microsoft.com/en-us/sustainability)."

The **Responsible Technology Hub** (RTH) helps in shaping the emerging technologies with a focus on youth and young professionals to connect together, collaborate and co-create a responsible technological future (responsibletechhub.com).

Open-Emission-Factors-DB

Open-Emission-Factors-Database, on GitHub, is designed to be constantly updated by contributors across the spectrum of sustainability researchers and professionals. It is freely available to all and maintained and validated by the team at climatiq.io.

The diagram following illustrates CO_2 equivalent emission for flight travel from Stockholm to Chennai.

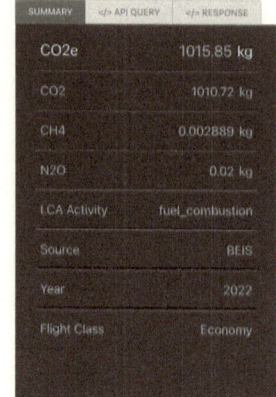

The diagram below illustrates emission factors for cloud computing in the CPU category.

Organizations Preparing for the Future

Many responsible companies have put programs in place to reduce their carbon footprint. Here are some accounts of their efforts I've gathered.

Shell

"EV Charging Stations, biofuels, the hydrogen transition, and chemicals are pillars of Shell's Climate plan. ... Shell is a massive business with more than 1 million commercial and industrial customers and about 30 million customers coming to its 46,000 retail service stations daily, according to the company's own estimates. The company organized its thinking around what it sees as growth opportunities, energy transition opportunities, and then the gradual obsolescence of its upstream drilling and petroleum production operations. In what it sees as areas for growth, Shell intends to invest around $5 billion to $6 billion in its initiatives, including the development of 500,000 electric vehicle charging locations by 2025 (up from 60,000 today) and an attendant boost in retail and service locations to facilitate charging." (Shieber 2021)

Schneider Electric building

"Research shows digital use cases can deliver up to 8% of greenhouse gas (GHG) reductions by 2050. ... IntenCity is a good example of this—the Schneider Electric building is equipped with internet of things (IoT)—enabled solutions, providing an end-to-end digital architecture that captures more than 60,000 data points every 10 minutes. It is smart-grid-ready and energy-autonomous, with 4,000 m2 of photovoltaic panels and two vertical wind turbines. IntenCity has its own building information modeling system, which is an exact reflection of the construction and energy model that is capable of reproducing the energy behavior of the real building." (George et al. 2022).

IKEA

Transforming into a circular business- IKEA is on an exciting journey of making more from less—designing all of their products to have circular capabilities, and aiming to use only renewable or recycled materials by 2030. As per IKEA,""We are committed to becoming a circular business and enabling our customers to live a more sustainable life. To make this a reality, one of our ambitions is to move towards the use of renewable and recycled materials by 2030, and to design products that are recyclable." (source:about.ikea.com)

Disney

"The Walt Disney Company has already established itself as a key player in the voluntary carbon market, and it intends to extend its offset purchasing program to address indirect emissions from its operations. As a result of Disney's double-digit internal carbon prices, the company has been able to pay above-average prices for new offset projects, particularly in the forestry sector. By the end of 2022, The Walt Disney Company is committed to having a science-based reduction goal in Scope 3 emissions, which include emissions from its products, service manufacturing, and delivery." (sustainabilitymag 2022)

Alphabet

"As Google's parent company and several former Google subsidiaries, Alphabet has a lengthy history of purchasing carbon offsets. Since 2020, the firm has purchased high-quality carbon offsets to offset Google's complete carbon footprint, including all operating emissions prior to the company becoming carbon neutral in 2007. This means that Google's net carbon impact for the course of its existence is now zero. Alphabet they are the first large corporation to accomplish this feat." (sustainabilitymag 2022)

Stitching It All Together

Global disruptions should be looked at as an opportunity to attract mindshare in a D-VUCAD world (disruption, volatility, uncertainty, complexity, ambiguity, and diversity) where we need to rebound with a stronger growth path.

What we need are some gigantic breakthrough digital use cases and solutions that are new and innovative, not overengineering but rather course-correcting existing solution, as need be, and let go or retire solutions that are not serving their purpose or would carry forward technological debts.

Governmental laws, regulations, and economic forums are enacted to pursue sustainable development goals. However, that doesn't stop us from thinking or contributing as individuals. It is vital for every person, society, and generational cohort to make attaining sustainable development goals a priority through the right digital technologies and smarter use cases. Try not to take the "boiling the ocean" kind of approach.

The change is already happening, and we all as individuals, societies, or organizations need to embrace that change in some form. The software engineering community plays a major role in this, where the responsible principles needs to be embedded within the lifecycle of the digital production value chain at all levels and measured with Objectives and Key Results (OKRs) and KPIs related to sustainability and human well-being. There is a vast variety of standards in place to ensure this change.

The agile way of product development induces more challenges in the incorporation and the impact on delivery. But the sustainability and well-being dimensions cannot be ignored, and we need to find ways of imparting them within teams while developing digital products. Some of those principles and other societal information discussed in this book can be leveraged and extended to the context of the appropriate software engineering.

Together, let's build and promote sustainability and human well-being for a greener future!

REFERENCES

Freitag, Charlotte; Berners-Lee, Mike; Widdicks, Kelly; Knowles, Bran; Blair, Gordon S.; Friday, Adrian (2022). "The real climate and transformative impact of ICT: A critique of estimates, trends, and regulations." *Patterns* 3, 8 (12 August), 100576, https://doi.org/10.1016/j.patter.2021.100340.

Gates, Bill (2021). *How to Avoid a Climate Disaster.* New York: Alfred A. Knopf.

George, Manju; O'Regan, Karen; Holst, Alexander (2022). "Digital solutions can reduce global emissions by up to 20%. Here's how." World Economic Forum Annual Meeting (23 May), https://www.weforum.org/agenda/2022/05/how-digital-solutions-can-reduce-global-emissions, accessed 17 February 2023.

Prof Dr Wim Vanderbauwhede"; "Low Carbon and Sustainable Computing" June 29, 2021 https://www.dcs.gla.ac.uk/~wim/

World Meteorological Organization (2022). *WMO Greenhouse Gas Bulletin* 14, https://public.wmo.int/en/media/press-release/more-bad-news-planet-greenhouse-gas-levels-hit-new-highs, accessed 26 October 2022

Climate change widespread, rapid, and intensifying – IPCC, https://www.ipcc.ch/2021/08/09/ar6-wg1-20210809-pr/, accessed August 9, 2021

United Nations, net-zero-coalition; https://www.un.org/en/climatechange/net-zero-coalition

Paris Agreement, 12 December 2015, https://unfccc.int/files/essential_background/convention/application/pdf/english_paris_agreement.pdf

Althor, G., Watson, J. & Fuller, R. Global mismatch between greenhouse gas emissions and the burden of climate change. *Sci Rep* **6**, 20281 (2016). https://doi.org/10.1038/srep20281

Malmodin, Jens & Bergmark, Pernilla. (2015). Exploring the effect of ICT solutions on GHG emissions in 2030. 10.2991/ict4s-env-15.2015.5. https://www.ericsson.com/en/reports-and-papers/research-papers/exploring-the-effects-of-ict-solutions-on-ghg-emissions-in-2030

Flinders University News (2022). 'Food Cyber Attacks Forecast,', 26 May, accessed 13 February 2023, https://news.flinders.edu.au/blog/2022/05/26/food-cyber-attacks-forecast.

"Measuring digital development: Facts and Figures 2022" https://www.itu.int/en/ITU-D/Statistics/Pages/facts/default.aspx

"9 things to know about Google's maps data: Beyond the Map"; https://cloud.google.com/blog/products/maps-platform/9-things-know-about-googles-maps-data-beyond-map

"Interesting eLearning Statistics You Need To Know"; March 3, 2023 ; https://colorwhistle.com/interesting-elearning-statistics/

Forbes (2018); https://www.forbes.com/sites/natalierobehmed/2018/12/03/highest-paid-youtube-stars-2018-markiplier-jake-paul-pewdiepie-and-more/?sh=11c637323d77

Globenewswire (October 31, 2022) ; https://www.globenewswire.com/news-release/2022/10/31/254

4834/0/en/Global-e-Commerce-Market-to-Hit-Sales-of-58-74-Trillion-By-2028-E-commerce-Market-Has-Come-a-Long-Way-Still-Need-to-Overcome-Some-Challenges.html

Nissan Motor Corporation (Feb 26 2019), https://global.nissannews.com/ja-JP/releases/release-41181d20da4d17b77ed88d700817c540-businesses-need-smarter-tech-in-their-fleets-to-survive-e-commerce-boom?source=nng

United Nations, Sustainable Development Goals, https://sdgs.un.org/goals

Royal Society (03 December 2020), https://royalsociety.org/news/2020/12/digital-tech-vital-net-zero-royal-society-report/

Medanta, 17-May-2022; https://www.medanta.org/patient-education-blog/what-is-the-memory-capacity-of-a-human-brain

H&M Group (2021). "Responsible AI, Is Better AI." Our Stories (15 June), https://hmgroup.com/our-stories/responsible-ai-is-better-ai, accessed 17 February 2023.

Wikipedia (2022). "Carbon Offset." Wikimedia Foundation. Last modified 27 February 2023, 10:27. https://en.wikipedia.org/wiki/Carbon_offset

How To Calculate Your Carbon Footprint (Oct 2013) https://www.3blmedia.com/news/how-calculate-your-carbon-footprint

Hope, Blaise (2022). "Top 10: Companies Committed to Reducing Carbon Footprint." Sustainability (17 May), https://sustainabilitymag.com/net-zero/top-10-companies-committed-to-reducing-carbon-footprint, accessed 17 February 2023.

PANCHA BHUTAS: YOGA'S 5 ELEMENTS OF NATURE https://www.yogabasics.com/learn/energy-anatomy/5-elements-of-nature/, accessed 2 Dec 2022

Top 10: companies committed to reducing carbon footprint (May 17, 2022); https://sustainabilitymag.com/net-zero/top-10-companies-committed-to-reducing-carbon-footprint

ICTFOOTPRINT.eu (2018). "ICT Standards." European Framework Initiative for Energy & Environmental Efficiency in the ICT Sector, https://ictfootprint.eu/en/ict-standards, accessed 17 February 2023.

INDUSTRIAL AI AND BETTER AND MORE SECURE SOCIETIES; https://www.qamcom.com/ai-into-better-and-more-secure-society/ accesed Dec 2 2022

Information Technology Industry Council (n.d.) "ITI View on Global ICT Standards." United States Information Technology Office, http://www.usito.org/news/key-ict-policy-principles-iti-view-global-ict-standards, accessed 17 February 2023.

National Science Foundation (2022). "Generating Renewable Hydrogen Fuel from the Sea." Research News (6 October), https://beta.nsf.gov/news/generating-renewable-hydrogen-fuel-sea, accessed 17 February 2023.

The NAU Review (2010). "New Study Finds Prescribed Burning Can Actually Reduce the US Carbon Footprint." Research & Academics (16 April), https://news.nau.edu/new-study-finds-prescribed-burning-can-actually-reduce-the-u-s-carbon-footprint/, accessed 17 February 2023.

Sadhguru (2021). "What are the Five Elements or Pancha Bhutas?" Isha (20 May). https://isha.sadhguru.org/us/en/wisdom/article/five-elements-pancha-bhuta, accessed 17 February 2023.

Shieber, Jonathan (2021). "EV Charging Stations, Biofuels, the Hydrogen Transition and Chemicals Are Pillars of Shell's Climate Plan." TechCrunch (February 11), https://techcrunch.com/2021/02/11/ev-charging-stations-biofuels-the-hydrogen-transition-and-chemicals-are-pillars-of-shells-climate-plan, accessed 17 February 2023.

Wong, Chloe (2022). "Top 8 Sustainable Tech Companies in the World Right Now." Earth.org (11 January), https://earth.org/sustainable-tech-companies, accessed 17 February 2023.

ACKNOWLEDGEMENTS

This work would not have been possible without a number of people who have given me contextual knowledge. I'm sincerely grateful, and many heartfelt thanks go out to them.

I would especially like to thank my colleagues, friends and leaders in the various organizations where I have worked to grow my career, shape my learning curve and solve real world problems.

I have executed solutions and consulted for more than fifty customers or enterprises across various industry segments, and I would also like to thank my clients who believed in me over the past eighteen years.

I have appreciated opportunities and support provided by the professors who educated me in engineering principles at Crescent Engineering College, which has hosted me as a key note speaker and guest lecturer as part of Alumni Connect; the many forums and societies that I am part of, including the World Economic Forum the Institute of Responsible Technology; my current employer as well as former employers in the IT industry; and all of the great people at AuthorHouse for the publication of this book.

Most especially, I would like to acknowledge my wonderful children, parents, and relatives for supporting me with love, good principles, and moments for creating this book. In particular, I thank my loving mother Thenmozhi, my beautiful wife Priya, and my dear sister Priyanka for their love and support throughout the book-writing process.

And finally, I thank the Almighty and Mother Nature for the blessings and support that make all things possible.

ABOUT THE AUTHOR

Balakarthik Baskaran is a technologist, philanthropist, digital transformation leader, and the strategic mind behind multiple digital transformations for a global clientele. He has more than eighteen years of experience in the software industry and has amazing insights on how technology can be used, produced, and curated responsibly.

www.ingramcontent.com/pod-product-compliance
Lightning Source LLC
Chambersburg PA
CBHW020452220526
45464CB00002B/957